Rap Your World

Rapping Rhymes about the Human Body

Thomas Kingsley Troupe

BLACK RABBIT BOOKS

Hi Jinx is published by Black Rabbit Books
P.O. Box 3263, Mankato, Minnesota, 56002.
www.blackrabbitbooks.com

Jen Besel, editor; Michael Sellner, designer;
Omay Ayres, photo researcher

Library of Congress Cataloging-in-Publication Data
Names: Troupe, Thomas Kingsley, author.
Title: Rapping rhymes about the human body / by Thomas Kingsley Troupe.
Description: Mankato, Minnesota : Black Rabbit Books, [2021] | Series: Hi Jinx. Rap your world | Includes bibliographical references. |
Audience: Ages 8-12 | Audience: Grades 4-6 |
Summary: "Combining musical rhythms and scientific information, students will explore the human body through poems meant for rapping"— Provided by publisher.
Identifiers: LCCN 2019031926 (print) |
LCCN 2019031927 (ebook) |
ISBN 9781623103255 (hardcover) |
ISBN 9781644664216 (paperback) |
ISBN 9781623104191 (ebook) |
Subjects: LCSH: Human body—Juvenile literature. |
Human physiology—Juvenile literature.
Classification: LCC QP37 .T718 2021 (print) |
LCC QP37 (ebook) | DDC 612—dc23
LC record available at https://lccn.loc.gov/2019031926
LC ebook record available at https://lccn.loc.gov/2019031927

Printed in the United States. 1/20

Image Credits

Dreamstime: Memo Angeles, 19; iStock: Adelevin, 4; Allevinatis, 20; Bella30127, 10, 23; Big_Ryan, 2–3; CandO_Designs, 17; exxorian, 6–7; Shutterstock: ADudkov, 3, 21; Alena Kozlova, 10–11; AnaBoo, 17; Arcady, 1; armi1961, Cover, 4–5; Christopher Hall, Cover, 3, 4–5, 7, 8, 10–11, 12, 15, 16, 19; Christos Georghiou, 20; ekler, 16; FrimuFilms, 18–19; Julien Tromeur, 14–15; Lightspring, 8–9; Lilu330, 15, 17; marity, 13; Memo Angeles, 1, 11, 13, 19; mohinimurti, 18; Mykhailo Bokovan, 8, 21; opicobello, 8–9; Pasko Maksim, 23, 24; pitju, 5, 8, 12, 21; Ron Dale, 5, 8, 12, 16, 20; Sergey Bogdanov, Cover; SlipFloa, Cover, 4; solar22, 18–19; STREET STYLE, Cover, 1, 4; studiovin, 16–17; totallypic, 11, 13, 18, 20; Tueris, Cover, 1, 19; Verzzh, 14–15;
Every effort has been made to contact copyright holders for material reproduced in this book. Any omissions will be rectified in subsequent printings if notice is given to the publisher.

Contents

Chapter 1

Rap Your World!

Yo, listen up readers. It's time to put down the phones.

It's time to talk about eyes, arms, organs, and bones.

But our chitchat won't be flat like a rug or doormat.

We're gonna rap about it all. What do you

think about that?!

The human body is amazing, as everyone knows,
From the color of your hair to your ugly little toes.
Deep inside we're full of blood, organs, and guts.
Outside we've got our hands, eyebrows, and butts.

We'll rap about our faces we all think are neat,
And our hands and our arms and our legs and our feet.
The organs in our bodies? We'll cover that too,
Turning rhymes in on time, like homework that's due!

Chapter 2

Hit It on the Head

Your head has some important parts. I'm not telling you lies.
But you wouldn't see a thing if it weren't for your eyes!
They come in lots of colors like green, brown, and blue.
Always active, eyes take in the world and keep things in view.

It's possible to blink five times in one second

Your sniffer's up next; that's what I call the nose.
It's the part that smells things, like old fish or a rose.
Our nose helps us breathe, and it works like a filter,
Keeping out **bacteria** that'll throw us **off-kilter**.

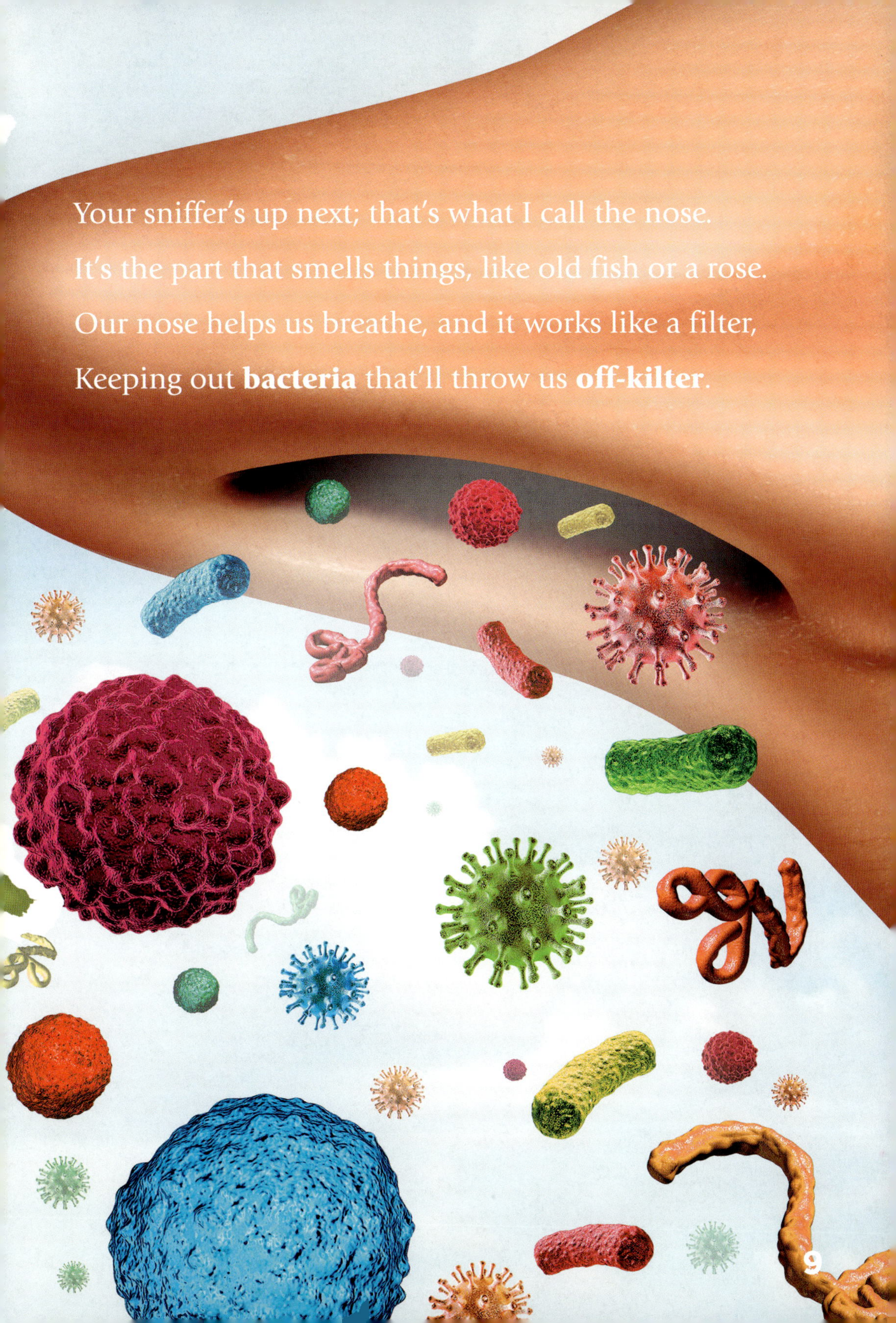

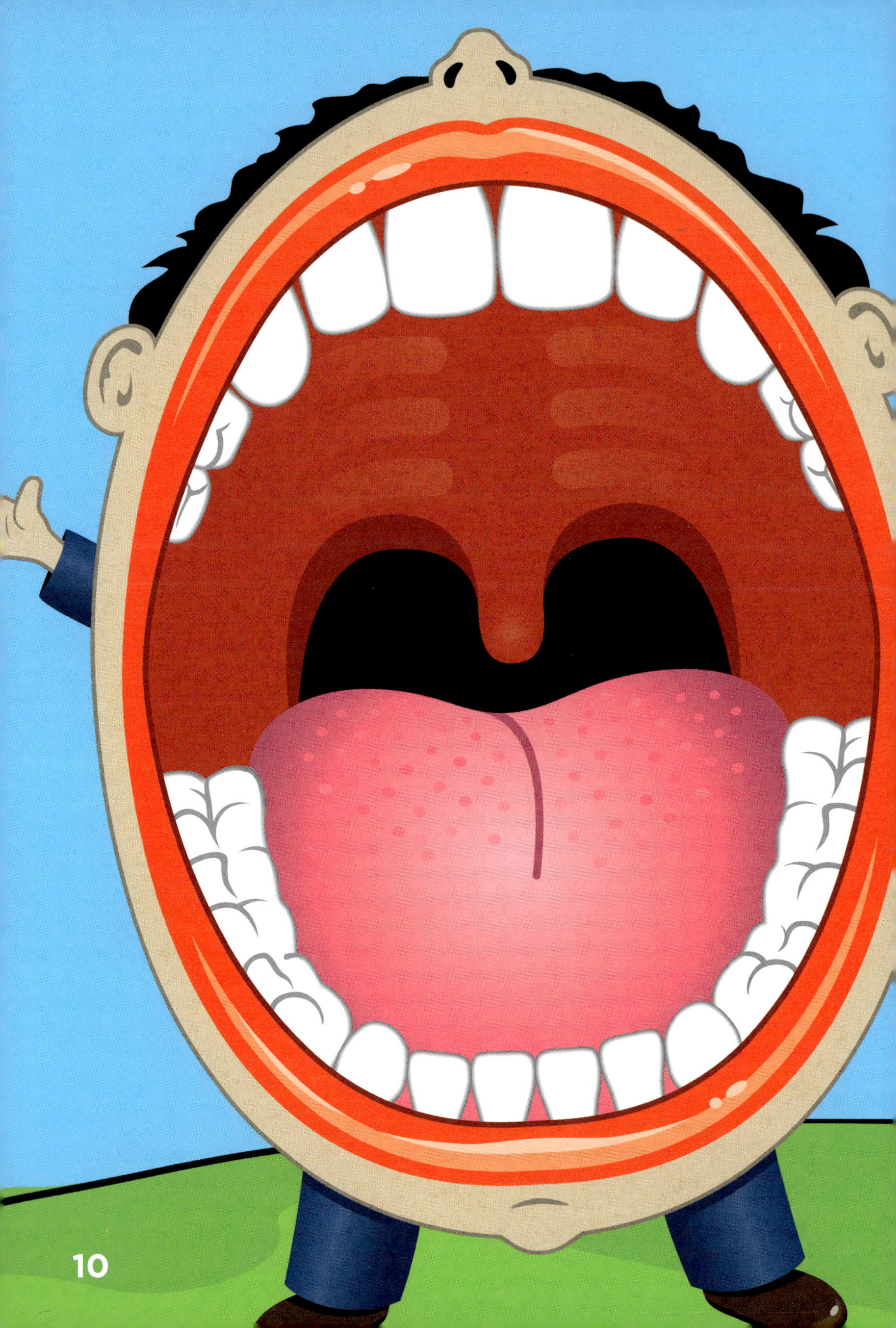

Fresh out of my mouth come the words that I speak.
I use mouthwash every day so my breath doesn't reek.
The mouth is the place where we taste all our food,
Full of teeth to chew it up, yeah I'm telling you, dude!

You know those things on the side of your head?
Your ears even work when you're asleep in your bed.
Ears help with balance and hearing your music,
But if the volume's too loud, then you know you could lose it.

Chapter 3

Hands and Feet

Yo, this part of the body makes lifting a charm.

The **appendages** I'm talking about? Your right and left arm!

Your arms bend at the elbows and are connected to shoulders.

They can help you lift and throw things, but maybe not boulders.

At the end of each arm, you should find a hand,

Good for clapping and waving and being part of a band.

Your hand's 10 fingers help you touch and pick up things,

And they're also great for sign language and wearing rings.

shoulder
elbow

Your two legs are made up of calves, knees, and thighs.
They keep you standing up, which should be no surprise.
Walk, run, hop, and maybe kick a soccer ball,
With the biggest bones in your body, your legs can do it all!

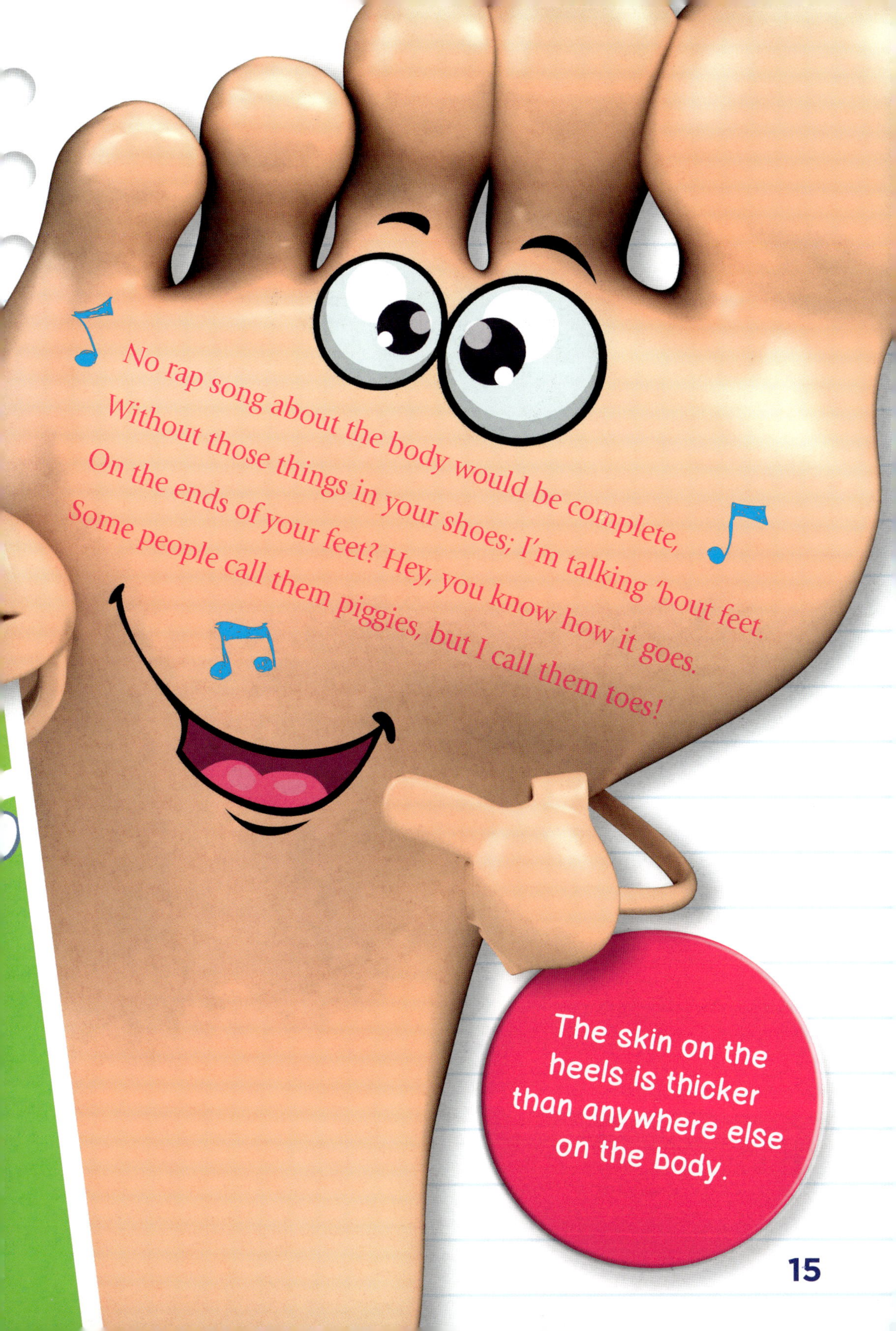

No rap song about the body would be complete,
Without those things in your shoes; I'm talking 'bout feet.
On the ends of your feet? Hey, you know how it goes.
Some people call them piggies, but I call them toes!

The skin on the heels is thicker than anywhere else on the body.

Chapter 4

The Insides

The inside of your body is where things start jumping.
Your heart is the organ that keeps the blood pumping.
With each beat, it pumps blood throughout the body,
Working all day and night, yo, that's not too **shoddy**!

There are two organs by your heart, but you might not believe,
They're your lungs—the organs that're helping you breathe,
Turning **oxygen** into energy to keep fueling your body.
Your nose helps keep your lungs clear; that's why it gets snotty!

Snot is like a filter. It keeps stuff in the air from going into the lungs.

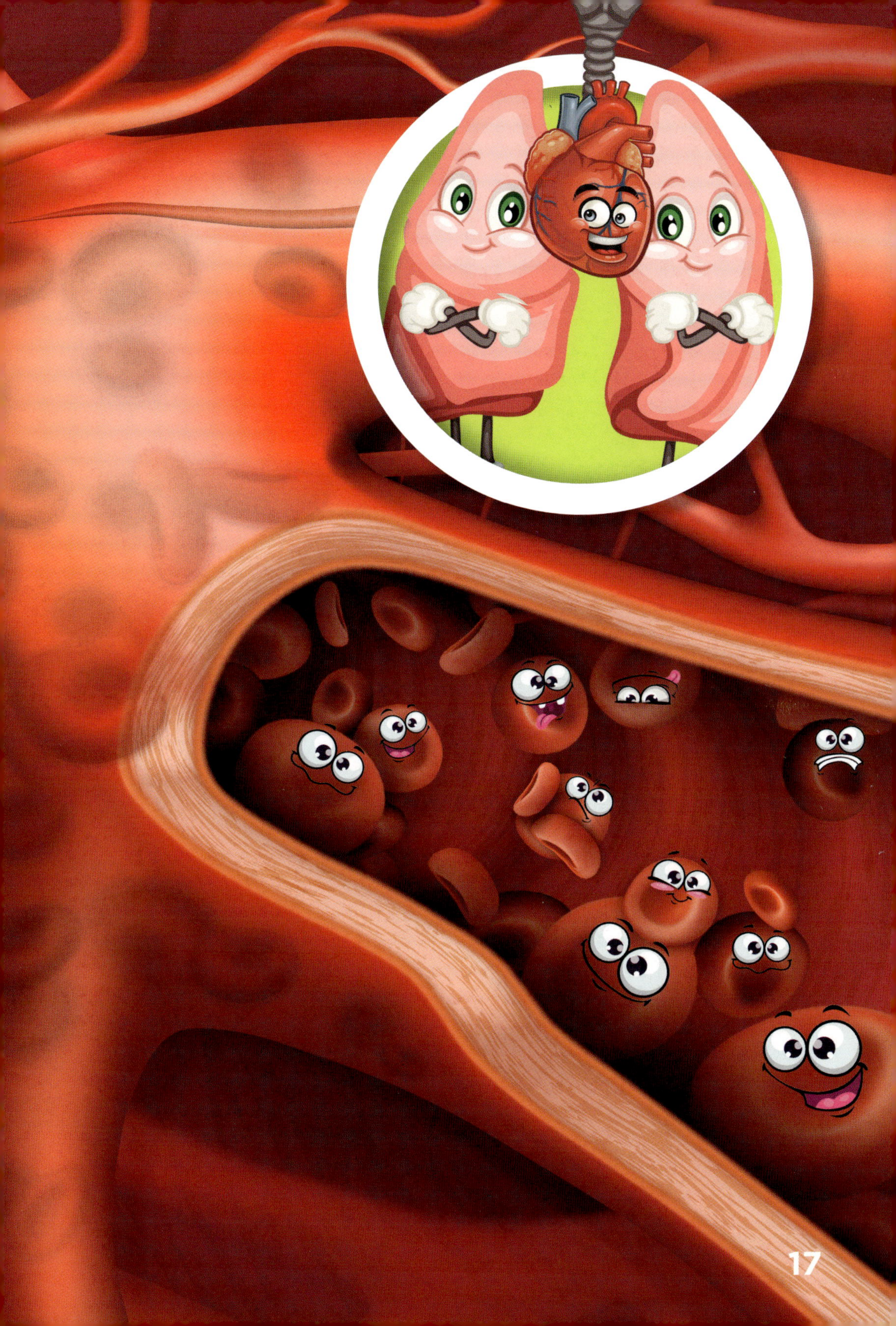

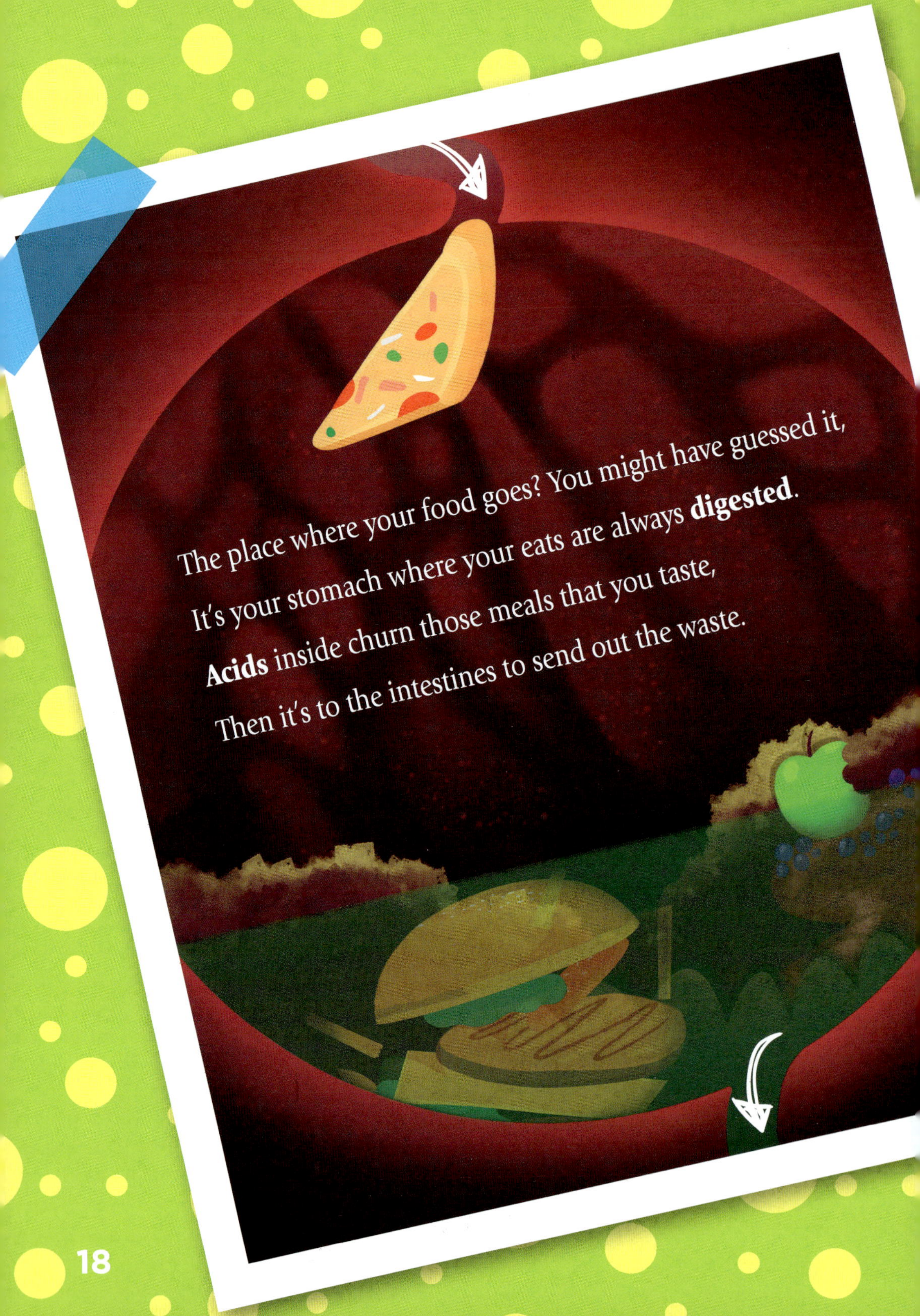

The place where your food goes? You might have guessed it,
It's your stomach where your eats are always **digested**.
Acids inside churn those meals that you taste,
Then it's to the intestines to send out the waste.

Your brain is the big organ sitting inside your head,
Running everything in your body so you don't end up dead.
Thinking and remembering, your brain's got memory,
And yo, it's how I always rhyme so cleverly!

Chapter 5

Get in on the HiJinx

What up? Think you can rap about the human body? First, list out a bunch of body parts. Find some weird ones, like the spleen! Next, look for cool facts about them. Try to find words that can be easily rhymed. Before long, you'll have a rap of your own.

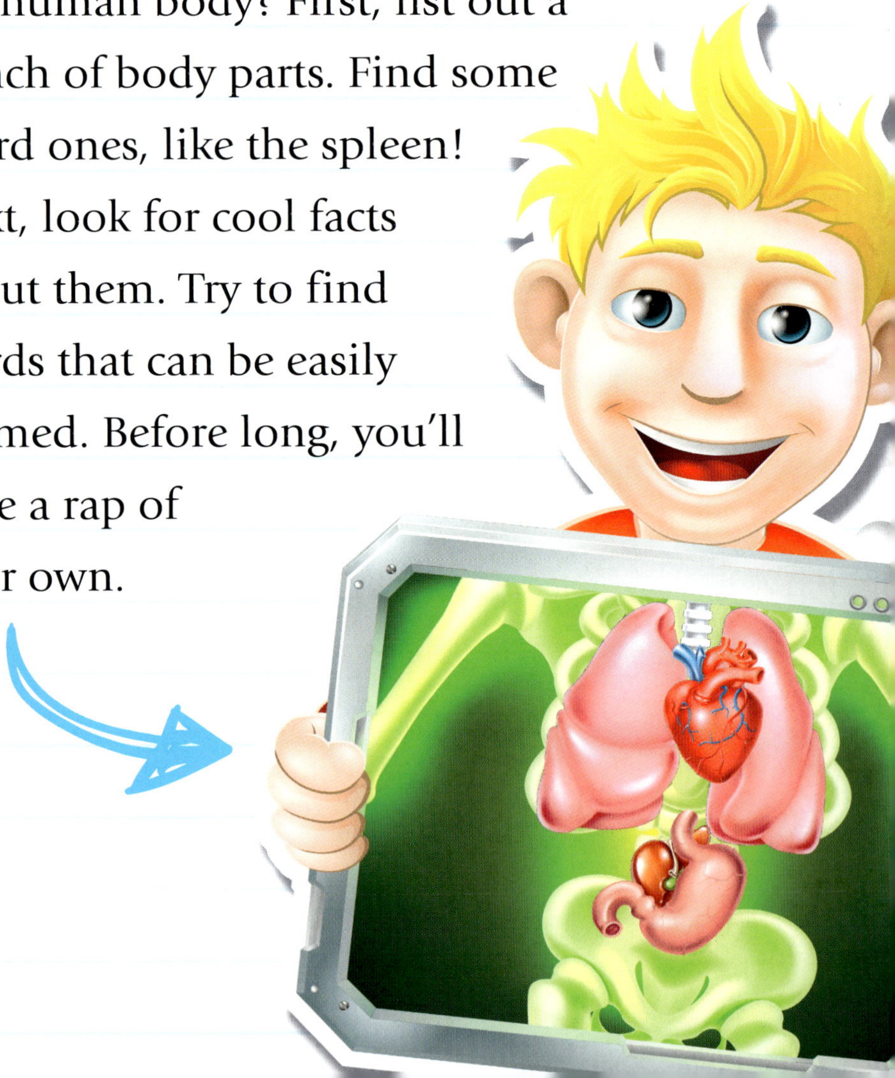

Take It One Step More

1. Did you read the lines to a beat? If you didn't, tap your hand on your leg in a steady rhythm. Try reading the words in time to the beat. How does that change your understanding of the information?

2. Rapping is a form of musical poetry. Is rapping a good way to learn information?

3. Have someone else read the raps out loud. Do they put **accents** in the same places you do?

GLOSSARY

accent (AK-sent)—an emphasis put on part of a word

acid (AHS-id)—a chemical substance that can dissolve some things

appendage (uh-PEN-dij)—a projecting part of a body, such as an arm or leg

bacteria (bak-TEER-ee-uh)—a small living thing

digest (DY-jest)—to change the food eaten into a form that can be used by the body

off-kilter (OFF KIL-tur)—not in perfect balance

oxygen (AHK-si-jin)—a gas that is necessary for life

shoddy (SHAW-dee)—hastily or poorly done

LEARN MORE

BOOKS

Braun, Eric. *Awesome, Disgusting, Unusual Facts about the Human Body.* Our Gross, Awesome World. Mankato, MN: Black Rabbit Books, 2018.

Minden, Cecilia, and Kate Roth. *Writing a Poem.* Write It Right. Ann Arbor, MI: Cherry Lake Publishing, 2019.

Pearson, Yvonne. *12 Great Tips on Writing Poetry.* Great Tips on Writing. Mankato, MN: 12-Story Library, 2017.

WEBSITES

Human Body Facts for Kids
www.scienceforkidsclub.com/human-body.html

Science for Kids: Bones and Human Skeleton
www.ducksters.com/science/bones.php

Writing a Rap – Getting Started
www.youtube.com/watch?v=o6NZoTqWLq4

TIPS FOR WRITING YOUR RAPS

Your rhymes should have a beat to them. Listen to music, and see if you can make your words fit with the beats.

For more traditional raps, each line should be about the same length. If you're rushing to fit all the words into a line, try cutting some.

Never be afraid to change your words. The best rappers in the world are constantly editing.